SpringerBriefs in Applied Sciences and Technology

SpringerBriefs present concise summaries of cutting-edge research and practical applications across a wide spectrum of fields. Featuring compact volumes of 50–125 pages, the series covers a range of content from professional to academic.

Typical publications can be:

- A timely report of state-of-the art methods
- An introduction to or a manual for the application of mathematical or computer techniques
- A bridge between new research results, as published in journal articles
- A snapshot of a hot or emerging topic
- An in-depth case study
- A presentation of core concepts that students must understand in order to make independent contributions

SpringerBriefs are characterized by fast, global electronic dissemination, standard publishing contracts, standardized manuscript preparation and formatting guidelines, and expedited production schedules.

On the one hand, **SpringerBriefs in Applied Sciences and Technology** are devoted to the publication of fundamentals and applications within the different classical engineering disciplines as well as in interdisciplinary fields that recently emerged between these areas. On the other hand, as the boundary separating fundamental research and applied technology is more and more dissolving, this series is particularly open to trans-disciplinary topics between fundamental science and engineering.

Indexed by EI-Compendex, SCOPUS and Springerlink.

More information about this series at http://www.springer.com/series/8884

Yuan Lu

Cell-Free Synthetic Biology

Yuan Lu
Department of Chemical Engineering
Institute of Biochemical Engineering
Tsinghua University
Beijing, China

Key Laboratory of Industrial Biocatalysis
Ministry of Education, Department of
Chemical Engineering
Tsinghua University
Beijing, China

ISSN 2191-530X   ISSN 2191-5318   (electronic)
SpringerBriefs in Applied Sciences and Technology
ISBN 978-981-13-1170-3   ISBN 978-981-13-1171-0   (eBook)
https://doi.org/10.1007/978-981-13-1171-0

This Springer imprint is published by the registered company Springer Nature Singapore Pte Ltd.
The registered company address is: 152 Beach Road, #21-01/04 Gateway East, Singapore 189721, Singapore

Preface

Synthetic biology faces many challenges due to unpredictability, complexity, incompatibility, and variability in the living cell systems. To break new ground with these challenges, an emerging approach cell-free synthetic biology (CFSB) has been adopted. Cell-free synthesis activates biological machinery in vitro without using living cells for engineering biology with more flexibility. CFSB has opened great opportunities and wide prospect for basic science research and health applications.

In the past decade, CFSB developed very fast. More and more researchers around the world take an active part in this exciting research field. All of these have catalyzed cutting-edge research in the synthesis of variable natural or unnatural biomolecules, the creation of artificial life, prototyping next-generation technologies, high-throughput workflow, and synthetic biomolecular network regulation. CFSB has transformed the studies of biological machinery in a profound and practical way for versatile applications.

In this book, we give an overview of emerging principles of CFSB and bioengineering, present how CFSB transforms life sciences research, and discuss how CFSB revolutionizes the environmental, biochemical, bioenergy, and human health industries. This book describes advanced studies in CFSB, emerging biotechnology that focuses on the development of different cell-free systems for fundamental and industrial research in areas such as synthesis of difficult-to-express and unnatural proteins, biosensing, artificial cells, and other emerging development trends. It is intended for students and researchers working in life sciences, synthetic biology, bioengineering, and chemical engineering.

Beijing, China   Yuan Lu
December 2018

Acknowledgements This work was supported by the National Natural Science Foundation of China (No. 21706144, 21878173) and the Beijing Natural Science Foundation (No. 2192023). This book can be completed thanks to the help of Xiaomei Lin, Peng Zhang, Wei Gao, Shengnan Ma, Qi Sun, Yingying Liu, Dong Liu, and Cheng-Yen Lin.

Contents

Chapter 1
An Introduction to Cell-Free Synthetic Biology

1.1 General Introduction

Synthetic biology, as a new life science discipline, involves structured construction of biological systems using modular and standardized engineering concepts. It enables engineering biological parts, devices, and systems for versatile applications such as medical therapeutics, medical diagnostics, bioenergy production, and understanding biology. Considering the complexity, variability, and redundancy of living cellular systems, a view comes from scientists who focus on the engineering of biosystems in vitro from the bottom up. It is like solution biochemistry for better applications and opens up a new understanding about biology. Therefore, an enabling technology called cell-free synthetic biology has been rapidly adopted and developed.

What is cell-free synthetic biology (CFSB)? It executes the central dogma of biology in an open or artificial environment without using the living cells (Fig. 1.1). Cell-free biosynthesis is exempt from the inherent constraints of cell-based methods. Because no living cells are involved, the living system environment can be freely adapted and engineered. The increased accessibility enables unprecedented freedom of design to engineering biological molecules and networks flexibly. Within a highly

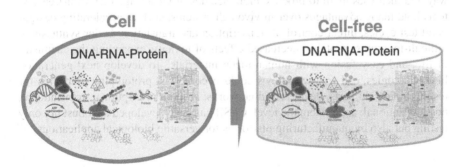

Fig. 1.1 Comparison of the cell-free system with the cell system

© The Author(s), under exclusive license to Springer Nature Singapore Pte Ltd. 2020
Y. Lu, *Cell-Free Synthetic Biology*, SpringerBriefs in Applied Sciences
and Technology, https://doi.org/10.1007/978-981-13-1171-0_1

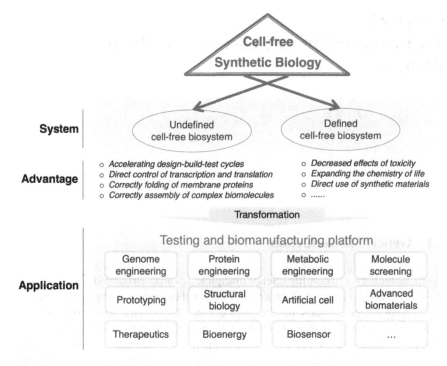

Fig. 1.2 Overview of cell-free synthetic biology (CFSB)

efficient design context, in vitro cell-free procedures have the prospect to complement in vivo cell-based efforts by rapidly providing the characterization data required for the rational biological design. In the cell-free biosystem, biomolecule production is uncoupled from the survival and reproduction need of the cell. Consequently, it is an ideal system if the product is toxic to the cell.

Cell-free biosynthesis method, rather than using living cells to make biomolecules, simply extracts the cells' natural biological machinery in an undefined or defined way and then uses them to produce biomolecules in vitro (Fig. 1.2). Cell-free systems hold many advantages over in vivo cell systems, such as accelerating design-build-test cycles, direct control of transcription and translation, better synthesis of difficult-to-express proteins, decreased effects of toxicity, expanding the chemistry of life, and easy fusion with human-made materials. To develop next-generation biotechnologies, cell-free workflows could become fast prototyping platforms for testing biological parts, devices, and networks, at the transcriptional, translational, posttranslational, and systematic level. CFSB has been developed as robust not only testing but also biomanufacturing platforms for versatile biological applications.

1.2 Cell-Free Biosystems

Cell-free biosystem uses natural protein-making machinery to produce proteins in vitro. Protein synthesis is considered the most basic level of the hierarchical structure of synthetic biology. Cell-free biosynthesis is the quickest way to obtain proteins from genes by transcription and translation. The representative breakthrough work using cell-free system is to decipher the genetic code done by Heinrich Matthaei and Marshall Nirenberg in 1961 [1], for which Nirenberg shared the 1968 Nobel Prize in Physiology or Medicine. Under the influence of Nirenberg's work, many scientists demonstrated the DNA-directed synthesis of polypeptides or proteins by using cell-free protein biosynthesis systems, and as a result, the 1970s become the first golden age for cell-free biology [2]. Since then, cell-free biotechnology has been a fundamental research tool for understanding biology, but it is challenging to industrialize because of high cost, low protein yield, and no scale-up technology. Until after 2000, many thanks to the efforts by some scientists and companies [3, 4], the challenges regarding cost, yield and scalability have been well addressed. The industrial cell-free protein production can reach the 100-liter scale [4]. Up to now, two main types of cell-free biosynthesis systems have been developed, including undefined and defined.

1.2.1 Undefined Cell-Free Biosystems

Undefined cell-free biosystems are crude cell extract-based systems. The crude extract holds primary transcription and translation functions. To ensure high-yield protein synthesis, besides the crude extract and DNA template, the system needs to be supplemented with other components, including RNA polymerase, energy-providing substrates, amino acids, NTPs, tRNAs, cofactors, and salts. The first thing to do for producing bioactive proteins with high yields using cell-free biosystem is choosing a proper chassis cell as the source of extracts [5]. The main factors to be considered are protein origin and complexity, posttranslational modification, application objective, and cost. Crude extracts have been successfully made from many different prokaryotic and eukaryotic cells. Prokaryotic cell-free systems include *Escherichia coli* [6] and *Bacillus subtilis* [7]. Eukaryotic cell-free systems include *Saccharomyces cerevisiae* [8], wheat germ [9], rabbit reticulocyte [10], and Chinese hamster ovary (CHO, mammalian) [11].

The most commonly used system is the *E. coli* extract-based cell-free system. The general workflow of *E. coli* extract-based cell-free reactions is shown in Fig. 1.3. The preparation of cell extracts is crucial for high translation efficiency, which includes harvesting cells at logarithmic growth phase and choosing the proper cell lysis approach. To keep the biological activities, ATP is required in cell-free reactions. Three kinds of chemicals with high-energy phosphate bonds, such as phosphoenolpyruvate (PEP), acetyl-phosphate, or creatine phosphate, are used for the ATP

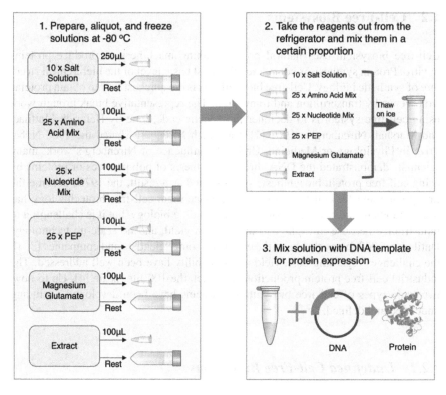

Fig. 1.3 The general workflow of *E. coli* extract-based cell-free reactions

generation, in combination with enzymes in the cell extracts such as pyruvate kinase, acetate kinase, or creatine kinase [12]. Some salts including glutamate salts, spermidine, and putrescine are essential for the proper function of biological processes. The cationic magnesium concentration always needs to be optimized for different systems. The cell-free reaction temperature and time also need to be optimized for different proteins.

1.2.2   Defined Cell-Free Biosystems

The most representative defined cell-free biosystem is *E. coli*-based PURE (Protein synthesis Using Recombinant Elements) system [13], which holds some advantages over cell extract-based system. In the undefined cell-free biosystems, the most significant disadvantage is those unclear components in the extract causes low reliability and stability. The degradation enzymes in the extract often cause the degradation problems of RNAs or proteins. In the PURE system, the components are clear and can be rationally manipulated. However, the disadvantage of the PURE system is high cost. Therefore, the PURE system is often used as a testing platform.

The defined *E. coli*-based PURE system is composed of purified ribosomes, tRNAs, aminoacyl-tRNA synthetases, translation factors, and several other essential enzymes for the accomplishment of protein synthesis [14]. The ribosome is the key translation machinery. Translation factors play essential roles in the translation process of PURE system, including three initiation factors (IF1, IF2, and IF3) binding to the ribosome during the translation initiation, three elongation factors (EF-G, EF-Tu, and EF-Ts) assisting the whole translation process, three release factors (RF1, RF2, and RF3) allowing for the translation termination, and ribosome recycling factor (RRF) for recycling the ribosome after the translation completion. Additionally, three other reactions are also essential for the protein synthesis, including the synthesis of mRNA by transcription, aminoacylation of tRNAs, and regeneration of the energy source. Therefore, T7 RNA polymerase, 20 aminoacyl-tRNA synthetases, creatine kinase, pyrophosphatase, myokinase, and nucleoside-diphosphate kinase are also formulated in the PURE system. Other reaction components including salt buffers, tRNA mixtures, 20 amino acids, and 4 nucleoside triphosphates are supplemented when cell-free reactions are performed [15].

1.2.3 Portable Freeze-Dried Cell-Free Biosystems

The dependence on living cells to manipulate genetic materials for biosynthesis of biopharmaceutical proteins faces many questions, such as biosafety concerns and requirement of specialized biotechnology skills, which limits the development speed of biotechnology and the application in developing regions. Biopharmaceutical proteins such as vaccines and antibodies must be distributed globally from centralized biomanufacturing factories, which usually requires the cold chain transportation for keeping the protein stability. All of these limitations above affect distribution costs and feature the challenge of delivering the advantages and benefits of these technologies and products to developing regions [16]. Designing and using freeze-dried cell-free biosynthesis systems could be an ideal choice to solve those challenges.

Freeze-drying the components of the cell-free system could reduce the storage volume and provide a longer shelf life at room temperatures. Bradley Bundy group has found that lyophilized extract with sucrose could maintain about 20% of the protein synthesis viability after 90 days of storage at room temperature [17]. In contrast, the extract in solution format only could retain less than 2% of synthesis viability after 30 days of storage at room temperature. The portable freeze-drying cell-free system has been successfully tested for the production of vaccines, peptides, and chemical drugs [16]. The freeze-drying cell-free biosystems also have been held on the filter paper as easy-to-use paper-based cell-free biosensors for prototyping the rapid detection of Ebola virus and Zika virus [18, 19].

References

1. J.H. Matthaei, M.W. Nirenberg, Characteristics and stabilization of DNAase-sensitive protein synthesis in *E. coli* extracts. Proc. Natl. Acad. Sci. USA **47**, 1580–1588 (1961)
2. F. Villarreal, C. Tan, Cell-free systems in the new age of synthetic biology. Front. Chem. Sci. Eng., 1–8 (2017)
3. J. Swartz, Developing cell-free biology for industrial applications. J. Ind. Microbiol. Biotechnol. **33**(7), 476–485 (2006)
4. J.F. Zawada et al., Microscale to manufacturing scale-up of cell-free cytokine production—a new approach for shortening protein production development timelines. Biotechnol. Bioeng. **108**(7), 1570–1578 (2011)
5. E.D. Carlson et al., Cell-free protein synthesis: applications come of age. Biotechnol. Adv. **30**(5), 1185–1194 (2012)
6. R. Ninomiya et al., Role of disulfide bond isomerase DsbC, calcium ions, and hemin in cell-free protein synthesis of active manganese peroxidase isolated from Phanerochaete chrysosporium. J. Biosci. Bioeng. **117**(5), 652–657 (2014)
7. R. Kelwick et al., Development of a Bacillus subtilis cell-free transcription-translation system for prototyping regulatory elements. Metab. Eng. **38**, 370–381 (2016)
8. R. Gan, M.C. Jewett, A combined cell-free transcription-translation system from Saccharomyces cerevisiae for rapid and robust protein synthe. Biotechnol. J. **9**(5), 641–651 (2014)
9. T.U. Arumugam et al., Application of wheat germ cell-free protein expression system for novel malaria vaccine candidate discovery. Expert Rev Vaccines **13**(1), 75–85 (2014)
10. J.A. Douthwaite, Eukaryotic ribosome display selection using rabbit reticulocyte lysate. Methods Mol. Biol. **805**, 45–57 (2012)
11. A.K. Brodel, A. Sonnabend, S. Kubick, Cell-free protein expression based on extracts from CHO cells. Biotechnol. Bioeng. **111**(1), 25–36 (2014)
12. M.C. Jewett, J.R. Swartz, Mimicking the *Escherichia coli* cytoplasmic environment activates long-lived and efficient cell-free protein synthesis. Biotechnol. Bioeng. **86**(1), 19–26 (2004)
13. Y. Shimizu et al., Cell-free translation reconstituted with purified components. Nat. Biotechnol. **19**(8), 751–755 (2001)
14. Y. Kuruma, T. Ueda, The PURE system for the cell-free synthesis of membrane proteins. Nat. Protoc. **10**(9), 1328–1344 (2015)
15. Y. Shimizu, T. Kanamori, T. Ueda, Protein synthesis by pure translation systems. Methods **36**(3), 299–304 (2005)
16. K. Pardee et al., Portable, on-demand biomolecular manufacturing. Cell **167**(1), 248–259 (2016)
17. M.T. Smith et al., Lyophilized *Escherichia coli*-based cell-free systems for robust, high-density, long-term storage. Biotechniques **56**(4), 186–193 (2014)
18. K. Pardee et al., Paper-based synthetic gene networks. Cell **159**(4), 940–954 (2014)
19. K. Pardee et al., Rapid, low-cost detection of Zika virus using programmable biomolecular components. Cell **165**(5), 1255–1266 (2016)

Chapter 2
Cell-Free Natural Protein Synthesis

Because no living cells are involved, the open cell-free biosystems can be flexibly regulated to produce proteins in a few hours. In the open system, the transcription and translation process can be directly controlled on demand by adjusting the temperature, salts, and redox environment and adding nature or human-made materials. Therefore, cell-free biosystem could be well adopted to produce proteins which are difficult to synthesize in living cells, such as membrane proteins and toxic proteins.

2.1 Membrane Protein

Representing about 30% of the proteins of an organism, membrane proteins are playing a significant role in various cell structures, as well as related to many biochemical and physiological processes [1]. They are anchored in the cell membrane for substance transportation, signal transmission, biological reactions as well as cell membrane shape maintenance. In the past few decades, many cases have confirmed to be correlated with diseases, as more than 50% of all drug target [2]. However, only a few numbers of membrane proteins are exogenously expressed properly for functional and structural characterizing, for the challenges of the specific chemical conditions of the cell membrane which naturally supporting the correct folding and function performing. Moreover, the membrane proteins usually are somewhat in a small amount within living cells, which makes the biosynthesis and purification limited in the laboratory along with the high cost for sufficient quantity production [2]. Furthermore, overexpression of them often brings about the cell toxicity because of the over uploading, which influences the fluidity of membrane and leading to cell lysis. Many efforts have made for decades to address these challenges, including mimicking hydrophobicity of natural phospholipid bilayer, optimizing the expression conditions, screening appropriate host, and modifying the translocation site of membrane proteins. In the cell-free biosystems, the capacities of direct condition controlling and substrate supplementing provide an attractive alternative to

© The Author(s), under exclusive license to Springer Nature Singapore Pte Ltd. 2020
Y. Lu, *Cell-Free Synthetic Biology*, SpringerBriefs in Applied Sciences
and Technology, https://doi.org/10.1007/978-981-13-1171-0_2

overexpress abundant functional membrane proteins for the fundamental research and pharmaceutical manufacturing.

Membrane proteins have been well adapted in cell-free biosystems assisted with detergent, liposome, and nanodisc (Fig. 2.1). Detergent can be added into the reaction environment to form an artificial hydrophobic environment to make membrane proteins soluble when they are initial peptide [3]. Many detergents have been exploited [4], such as Brij derivatives, Tween derivatives, DDM, Triton X-100, and so on. Lipids are added to mimic the natural phospholipid bilayer, allowing membrane proteins inserted into the bilayers directly, which has the advantage of functional or structural studies for the similarly natural folding state. Some studies have verified the improvement of the transcription and translation efficiency with the supplement of liposome [4]. Notably, the combination of detergent and lipids also shows their possibility of the improvement in functional cell-free membrane protein expression [4]. Nanodiscs are recently employed as an addition in the cell-free reaction. In this mode, membrane proteins are inserted into nanodiscs, which consist of lipid bilayer patches bound with a belt of membrane scaffold proteins [3].

Cell-free protein synthesis systems have been frequently employed for functional membrane protein expression either for fundamental research or pharmaceutical applications. The typical transmembrane G protein-coupled receptors are very difficult to be further studied because of their structural complexity and lack of appropriate screening systems, resulting in that many GPCRs are still olfactory receptors having urgent requirements of odorants screening. To address this, cell-free synthesis systems offer an elegant route for their expression and screening [5]. The gene templates are derived from the database, supplied into the transcription-translation

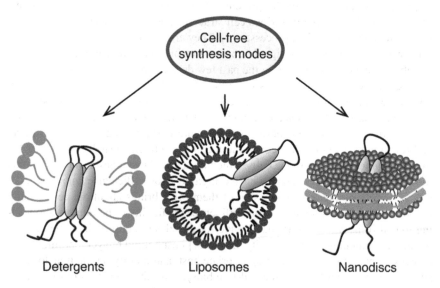

Fig. 2.1 The membrane protein synthesis in cell-free biosystems with detergents, liposomes, or nanodiscs

coupled cell-free biosynthesis systems. In this case, protein-protein interaction of GPCRs can be easily observed in cell-free biosystems supplemented with lipids.

2.2 Antibody

Monoclonal antibodies (mAbs), as the high specificity and affinity toward the target, have become a powerful tool in the clinical application of tumor treatment, autoimmune as well as infectious disease treatment, and are employed for biotechnological research. In particular, therapeutic mAbs has been continuously adapted for commercial availability along with the development of hybridoma technologies for functional mAb production [6]. However, the mammalian cell expression systems are still the dominant platform for industrial function mAb production, which is expensive and time-consuming. The prokaryotic expression systems, such as *E. coli* and *B. subtilis*, are currently developed for sufficient expression of complex proteins at high yields. Although many signs of progress have been made for the adaption of prokaryotic expressions, such as the development of single-chain variable fragment (ScFv), fragment of antigen binding (Fab), single-domain antibody (sdAb) and small size nanobody, the lack of modification ability for proper protein folding still limits their applications. Cell-free biosystems have great potential to solve the challenges on antibody synthesis, such as disulfide-bonded formation and high-throughput mAb generation [6].

Cell-free expression biosystem has its advantages for mAb design and synthesis. As is known to all, mAbs have the secondary structure of disulfide bonds, which limits the applications of bacterial systems. It is easy to supply the redox buffer and isomerases (DsbA, DsbC, FkpA, and SurA) in the cell-free system to help the formation of disulfide bonds [7]. The eukaryote with post-translation systems is more favorable for the mAb synthesis, such as wheat germ, rabbit reticulocytes, insect cells, and Chinese Hamster Ovary (CHO) cells. Although the eukaryote-based cell-free systems succeed in mAb synthesis, the cost and scale are still the limitations to get adequate mAbs for laboratory research and commercial manufacturing. Despite this, many efforts are still made for bacterial systems for the low-cost, large-scale and high-yield production [7]. It is demonstrated that cell-free synthesis is a valid option for the generation and assay of cell-penetrating antibodies [7]. Various antibodies, such as Fab that has the affinity to Botulinum neurotoxin serotype B and cytotransmab T Mab4, were successfully expressed in modified *E. coli* cell-free systems [8]. Furthermore, not only functional antibodies fragments are produced in the cell-free systems, but also it helps to elucidate the structural basis of EGFR-ECD recognition by the antibody [9].

2.3 Toxic Protein

Overexpression of proteins closely related to the life cycle or cell apoptosis sometimes is toxic to cells. These proteins commonly reach their maximal level for cyto-
toxicity to maintain the biological balances or host toxicity. More importantly, lots of
toxic proteins are relative to critical diseases or physiological process. So there is an
urgent need for a platform that toxic proteins could be overexpressed in an efficient
and simple way. Therefore, as the advantage of bypassing the physiological activities
of the living cell, cell-free biosynthesis systems have been employed as a powerful
tool to address the challenges of expression and characterization of toxic functional
proteins in vitro [10]. Many cases have confirmed that cell-free biosynthesis systems
are more superior for higher yield production of toxic proteins than cellular production [11]. Involved in apoptosis, the cytotoxic protein identified in cabbage butterfly,
named pierisin-1, were also successfully produced in a cell-free system with high
catalytic activity [10]. The toxin TDH, considered as a pore-forming toxin, was
reported that it could be expressed in significant amounts in the E. coli-based cell-
free biosystem, and the yields achieved almost 100 fold higher than previous cellular
expression [12].

2.4 Polypeptide

Biological production of bioactive polypeptides has attracted more and more atten-
tion over the past few decades. As the characteristics of small molecular weight
and single structure units, polypeptides used to be synthesized in chemical reactions with amino acid condensation, achieved by steps of blocking and deblocking,
which is expensive, time-consuming and not flexible enough for the wide variety of
polypeptides. Currently, many efforts have been made on biological synthesis, for the
possibilities to control the final product only with the gene sequence manipulation.
At the same time, many polypeptides are likely to cause toxicity to the host and need
a high-throughput method for screening and characterizing. As simple translation
products, polypeptides have already been expressed in the cell-free systems in the
early system development. Fused with GFP, an antibacterial polypeptide cecropin
P1 were synthesized in E. coli-based cell-free system [13]. Furthermore, expressing
elastin-like polypeptides (ELPs) with the structure of the repeating pentapeptides
(Val-Pro-Xaa-Gly) in the cell-free systems also have been demonstrated [14].

2.5 Enzyme

Enzymes act as biocatalysts facilitating diverse reactions in the organism. The lacking or mutations of enzymes usually raise the problems of severe diseases, such as
cancer, arthritis. Hence, this highlights the need to characterize them specifically.

Many advantages can be marked in cell-free enzyme synthesis systems, including their ability to express proteins that are difficult or toxic to express inside the cell, easily labeling for downstream detection, allowing for high-throughput screening and time-saving. Also, the storage stability issue and high-throughput screening limitation can be greatly reduced by the combination of synthesis and analysis of the target protein [15]. It has been demonstrated that active kinases can be synthesized by the cooperation between cell-free synthesis systems and immobilization technologies bypassing the step of purification [16]. Moreover, the advantage of condition controlling makes cell-free systems well suitable for high-yield protein expression [17]. Furthermore, clinical enzymes can be functionally synthesized with high efficiency in the eukaryotic cell-free system with robust post-modification capability [18].

References

1. L. Liguori, B. Marques, J.L. Lenormand, A bacterial cell-free expression system to produce membrane proteins and proteoliposomes: from cDNA to functional assay. Curr. Protoc. Protein. Sci. **Chapter 5**, Unit 5 22 (2008)
2. R.B. Quast et al., Automated production of functional membrane proteins using eukaryotic cell-free translation systems. J. Biotechnol. **203**, 45–53 (2015)
3. S. Ruehrer, H. Michel, Exploiting Leishmania tarentolae cell-free extracts for the synthesis of human solute carriers. Mol. Membr. Biol. **30**(4), 288–302 (2013)
4. F. Junge et al., Advances in cell-free protein synthesis for the functional and structural analysis of membrane proteins. N. Biotechnol. **28**(3), 262–271 (2011)
5. S. Ritz et al., Cell-free expression of a mammalian olfactory receptor and unidirectional insertion into small unilamellar vesicles (SUVs). Biochimie **95**(10), 1909–1916 (2013)
6. T. Ojima-Kato et al., 'Zipbody' leucine zipper-fused Fab in *E. coli* in vitro and in vivo expression systems. Protein Eng. Des. Sel. **29**(4), 149–157 (2016)
7. S.E. Min et al., Cell-free production and streamlined assay of cytosol-penetrating antibodies. Biotechnol. Bioeng. **113**(10), 2107–2112 (2016)
8. I.S. Oh et al., Cell-free production of functional antibody fragments. Bioprocess Biosyst. Eng. **33**(1), 127–132 (2010)
9. T. Matsuda et al., Cell-free synthesis of functional antibody fragments to provide a structural basis for antibody-antigen interaction. PLoS One **13**(2), e0193158 (2018)
10. J.H. Orth et al., Cell-free synthesis and characterization of a novel cytotoxic pierisin-like protein from the cabbage butterfly Pieris rapae. Toxicon **57**(2), 199–207 (2011)
11. A. Ranji et al., Chapter 15—transforming synthetic biology with cell-free systems, in *Synthetic Biology*, ed. by H. Zhao (Academic Press, Boston, 2013), pp. 277–301
12. S. Bechlars et al., Cell-free synthesis of functional thermostable direct hemolysins of Vibrio parahaemolyticus. Toxicon **76**, 132–142 (2013)
13. K.A. Martemyanov et al., Cell-free production of biologically active polypeptides: application to the synthesis of antibacterial peptide cecropin. Protein Expr. Purif. **21**(3), 456–461 (2001)
14. H.S. Chu et al., The effects of supplementing specific amino acids on the expression of elastin-like polypeptides (ELPs). Protein Expr. Purif. **74**(2), 298–303 (2010)
15. H. Singh et al., Application of the wheat-germ cell-free translation system to produce high temperature requirement A3 (HtrA3) proteases. Biotechniques **52**(1), 23–28 (2012)

16. D.M. Leippe et al., Cell-free expression of protein kinase a for rapid activity assays. Anal. Chem. Insights **5**, 25–36 (2010)
17. Y. Wang et al., Establishment and optimization of a wheat germ cell-free protein synthesis system and its application in venom kallikrein. Protein Expr. Purif. **84**(2), 173–180 (2012)
18. D.G. Mudeppa, P.K. Rathod, Expression of functional Plasmodium falciparum enzymes using a wheat germ cell-free system. Eukaryot. Cell **12**(12), 1653–1663 (2013)

Chapter 3
Cell-Free Unnatural Protein Synthesis

Naturally, proteins generally consist of twenty natural amino acids (NAAs) as building blocks, which can form a nearly unlimited number of combinations by random combination to realize structural and functional diversity. Unfortunately, it is not enough using these natural blocks to generate proteins with specific characteristics. Therefore, incorporating unnatural amino acids (UNAAs) with some novel groups to expand the repertoire of structures and functions of proteins is becoming more and more popular [1]. Diverse bioreactivity from novel functional groups are unreachable to natural proteins, which opens gates for new protein engineering and provides new ways for fundamental research, therapeutics, and synthetic biology [2]. Incorporation of UNAA at a specific site has been used for studying protein structure and dynamics, characterizing protein-protein interactions, mimicking post-translational modifications of proteins like eukaryote, and synthesize novel products hard to be created by other methods, such as enzymes, biomaterials, and therapeutics [3].

Up to now, more than 150 OTSs have been used to the incorporation of UNAAs containing novel functional groups at specific sites of proteins [4]. One representative study using cell-free biosystem to incorporate UNAA is from the Schultz group, in which UAA was first incorporated at an amber stop codon replacing a sense codon of a protein using *E. coli* extract system [5]. As a complement to cellular systems, cell-free biosystems offer some advantages. First, unnatural components can be flexibly added into the open cell-free synthesis biosystems at precise concentrations. Second, cell-free biosystems are not constrained by toxicity problems of unnatural substrates or products. Third, there is high unnatural protein yield because all resources in cell-free biosystems are used for the synthesis of target unnatural proteins. Fourth, almost all UNAA incorporation approaches work in cell-free biosystems. Cell-free biosystems have been developed as a robust platform for unnatural protein synthesis with versatile applications (Fig. 3.1).

Y. Lu, *Cell-Free Synthetic Biology*, SpringerBriefs in Applied Sciences and Technology, https://doi.org/10.1007/978-981-13-1171-0_3

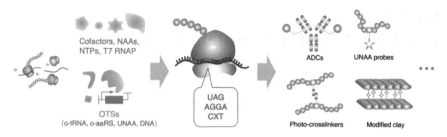

Fig. 3.1 Cell-free unnatural amino acid (UNAA) incorporation

3.1 UNAA Incorporation Approaches

3.1.1 Global Suppression

Global UNAA substitution generally utilizes auxotrophic strains, which cannot synthesize one particular NAA or several NAAs, substituted with structurally similar UNAAs [6]. The most popular examples include the replacement of methionine (Met) by azidohomoalanine (AHA) and homopropargylglycine (HPG) to introduce azide and alkyne groups [7]. An advantage of global substitution is that it can be accessible to incorporate the same UNAAs at multiple sites [8]. In the preparation of cell-free reactions, the first is to prepare extract derived from auxotrophic strains or deplete all NAAs in the extract [8]. Subsequently, supply required NAAs and UNAAs without natural counterparts of UNAAs into the reaction. In theory, global suppression can be used by combined with all codon reassignment approaches for UNAA incorporation to creating a promising future for further protein development with novel functions [9].

3.1.2 Stop Codon Suppression

Currently, stop codon suppression is the most widely used UNAA incorporation method. There are three termination codons in natural 64 codons to prevent the translation process recognized by release factors (RFs) [10]. However, certain organisms can expand genetic code by using the stop codon to encode particular amino acids [11]. This behavior is defined as stop codon suppression, which is mediated by suppressor tRNAs. By this method, selenocysteine [12] and pyrrolysine [12] are incorporated by opal codon and amber codon, respectively. Stop codon suppression, particularly mediated by the amber codon, has become one of the most widely used methods for site-specific UNAA incorporation. As RFs can compete the suppressor tRNAs with stop codons, there are two types of products that could be obtained: one is the full-length protein with UNAA, and the other one is truncated protein terminated at stop codon site [13]. Many efforts have been made for improving the suppression efficiency and increasing the yield of full-length protein containing UNAA. In this condition, the cell-free system has its unique advantages for UNAA incorporation due to its open environment. The cell-free reactions are flexibly regulated by adding or removing the components. For example, in prokaryotes, RF1 and RF2 respectively recognize the amber codon and the opal codon. Both RFs can recognize the ochre codon [14]. Based on this, a cell-free biosystem derived from RF1-depleted *E. coli* extract has been developed in which the suppression efficiency of the amber codon is significantly improved. Besides this, thermosensitive RF1 variants inactivated by heating and inhibition of RF1 by antibodies and aptamers can be realized in cell-free biosystems [15]. Recently, Hong and coworkers have demonstrated that incorporating multiple UNAAs could be achieved by stop codon suppression based on extract derived from a genetically-recoded *E. coli* lacking RF1 [16].

3.1.3   Frame-Shift Codon Suppression

The utilization of frame-shift codons is based on a series of tRNAs containing an extended anticodon loop resulting in decoding codons with more than three nucleobases [17]. In this approach, the codons with four or five bases have been successfully used for incorporating UNAAs in cell-free biosystems [18]. Particularly, the quadruplets derived from rarely used codons can efficiently compete with endogenous tRNAs decoding corresponding triplet codon [19]. By particular design, when the triplet codon is decoded at the quadruplet site, the translation process is terminated. It is beneficial to determine that the product is the frame-shift protein [20]. Using *E. coli* CFPS system, Sisido realized UNAAs incorporation at higher efficiency and multiple UNAA incorporations by frame-shift codon suppression comparing with amber suppression. Furthermore, frame-shift codon suppression has also been applied to eukaryotic cell-free biosystems, such as *Sf21* extract and rabbit reticulocyte extract-based systems [21].

3.1.4   Sense Codon Reassignment

Because of the degeneracy of codons, it is enough to encode an organism with 30–40 sense codons [12]. Therefore, there are more than 20 redundant codons that can be reassigned to other new amino acids. Sometimes, this process can be naturally taken place. For example, the CUN (N represents A, U, G or C) codons which encode leucine are rewritten to threonine in yeast mitochondria. There are only a few mutations to tRNAs and aaRSs for forming the new orthogonal pairs used in the rewriting process. In recent research, mutations have been made at the peptidyl transferase center of the 50S subunit of the ribosome to prevent it from utilizing several natural tRNAs during translation [12]. Especially, rare codons are the promising candidates used for genetic code expansion [22]. Among currently used cell-free systems, the PURE system is wholly composed of several purified recombinant protein factors [10]. Therefore, some translational components can be easily regulated, such as amino acids, tRNAs, aaRSs and so on, when UNAAs are incorporated into proteins in *E. coli*-based CFPS system [23].

3.1.5 Unnatural Base Pair

The use of unnatural base pairs to generate unique codons for incorporating UNAA at a specific site is a promising approach compared to the above methods. The limitations of the strategies presented above include the number of available codons, the orthogonality of translational components, and potential influences on protein structure and function. The method using unnatural base pairs can create countless unique codons which efficiently encode UNAAs [24]. The advantages and limitations have been described in a recent review about unnatural base pairs [25]. In cell-free biosystems, the suppression of unnatural base pairs is efficient for the UNAA incorporation. However, much more efforts must be made to promote the applications of unnatural base pairs.

3.2 Applications of Unnatural Protein

3.2.1 Biophysical Probes

Incorporating biophysical probes into proteins at specific sites, as a detection method with highly sensitive, can be employed for quantifications and structural studies without interfering protein function [26]. The proteins with stable isotope labeled can be produced in cell-free biosystem with the least costly amino acids. Therefore, cell-free biosystems represent a superior approach especially for FT-IR, NMR characterization and MS quantification [27]. In addition to $^{13}C/^{15}N$-labeled amino acids, ^{19}F-replaced UNAAs can be used for NMR studies with clean background [26]. ^{19}F, an important biophysical probe, has good sensitivity of chemical shift to the local environment [28]. Moreover, using heavy atom-replaced UNAAs can efficiently facilitate X-ray crystallographic structure determinations [29].

Fluorescent UNAA is also another valuable biophysical probes. Fluorescent UNAAs can be used for determining the protein yields, no need for radioactive labeling and liquid scintillation counting [30]. Incorporating fluorescent probes at specific sites to provide information about protein conformation, localization, and interactions [31]. Sisido and coworkers observed the changes in FRET signal of calmodulin variant after ligand binding by incorporating two different BODIPY derivatives with pre-charged frameshift-tRNAs [32]. Moreover, fluorescent UNAAs can be utilized for screening MMP-9 inhibitors by fluorescence correlation spectroscopy [33].

3.2.2 Enzyme Engineering

The traditional method to improve the function of an enzyme is substituting one NAA with one of the other 19 NAAs. However, when natural enzymes function as biocatalysts, it is difficult to maintain their stability and optimum activity under a tough condition, such as exposure to heat and organic solvents. However, genetically encoded UNAAs offer great promise for constructing artificial enzymes with novel activities [34]. The incorporation of fluorinated amino acids is significant for increasing stability. It is because that fluorination cannot only increase hydrophobicity to stabilize proteins but also closely protect the shape of the side chain. The stability and activity of enzymes derived from *Candida antarctica* have been enhanced by residue-specific fluorination of aromatic residues of lipase B [35]. Particularly, the incorporation of multiple UNAAs is promising for improving the characteristics of the structure and function of proteins. However, there are still two challenges in this method. First, the incorporation sites are often non-specific. Second, global incorporation of UNAAs usually interferes the process of protein folding, particularly for larger proteins [35].

3.2.3 Biopharmaceuticals

Over the past decade, enormous efforts have been made to optimize the efficacy and pharmacokinetics of protein therapeutics by protein engineering [36]. However, the structure and function of protein therapeutics are restricted by limited NAA building blocks. The protein therapeutics produced by traditional chemical modification often is a mixture of products in which the most is the undesired products. Moreover, due to this heterogeneity, it is hard to optimize the pharmacological properties of protein therapeutics further. The UNAA incorporation at a specific site of proteins by genetic expansion is a promising opportunity in the pharmaceutical industry. Past studies highlight the utility of cell-free UNAA incorporation biosystem for the production of novel vaccines and therapeutics. Moreover, it also can be used as an attractive tool for drug discovery. Swartz and coworkers incorporated UNAA in cell-free biosystem developing a novel pipeline for producing decorated virus-like particles which can be used as imaging agents and potential vaccines [37]. Sutro Biopharma company has synthesized the antibody-drug conjugates containing UNAA at the specific site [38]. Generally, there are two approaches to improve the half-life of therapeutic peptides and proteins. The one is the use of genetic fusion proteins, and the other one is non-specific conjugation of a synthetic polymer like polyethylene glycol (PEG) [39].

However, the products of randomly PEGylated are heterogeneous and contain species whose efficacy and in vivo pharmacokinetics (PK) is lower. Conversely, scanning the sites for conjugation by UNAA-based protein engineering helps obtain the ideal bioconjugates [40]. To date, several proteins PEGylated at the specific site have been synthesized via UNAA-based protein engineering to handle many diseases [41].

References

1. B.J. Des Soye et al., Repurposing the translation apparatus for synthetic biology. Curr. Opin. Chem. Biol. **28**, 83–90 (2015)
2. L. Wang, Genetically encoding new bioreactivity. N. Biotechnol. **38**(Pt A), 16–25 (2017)
3. S.H. Hong, Y.C. Kwon, M.C. Jewett, Non-standard amino acid incorporation into proteins using *Escherichia coli* cell-free protein synthesis. Front. Chem. **2**, 34 (2014)
4. P. O'Donoghue et al., Upgrading protein synthesis for synthetic biology. Nat. Chem. Biol. **9**(10), 594–598 (2013)
5. C.J. Noren et al., A general-method for site-specific incorporation of unnatural amino-acids into proteins. Science **244**(4901), 182–188 (1989)
6. A. Dumas et al., Designing logical codon reassignment—expanding the chemistry in biology. Chem. Sci. **6**(1), 50–69 (2015)
7. Y. Lu et al., *Escherichia coli*-based cell free production of flagellin and ordered flagellin display on virus-like particles. Biotechnol. Bioeng. **110**(8), 2073–2085 (2013)
8. L. Merkel et al., Parallel incorporation of different fluorinated amino acids: on the way to "teflon" proteins. ChemBioChem **11**(11), 1505–1507 (2010)
9. T.H. Yoo, A.J. Link, D.A. Tirrell, Evolution of a fluorinated green fluorescent protein. Proc. Natl. Acad. Sci. USA **104**(35), 13887–13890 (2007)
10. R.B. Quast et al., Cotranslational incorporation of non-standard amino acids using cell-free protein synthesis. FEBS Lett. **589**(15), 1703–1712 (2015)
11. H.S. Park et al., Expanding the genetic code of *Escherichia coli* with phosphoserine. Science **333**(6046), 1151–1154 (2011)
12. A. Böck et al., Selenoprotein synthesis: an expansion of the genetic code. Trends Biochem. Sci. **16**(12), 463–467 (1991)
13. M. Stech et al., Cell-free systems: functional modules for synthetic and chemical biology. Adv. Biochem. Eng. Biotechnol. **137**, 67–102 (2013)
14. E. Scolnick et al., Release factors differing in specificity for terminator codons. Proc. Natl. Acad. Sci. USA **61**(2), 768–774 (1968)
15. D.E. Agafonov et al., Efficient suppression of the amber codon in *E. coli* in vitro translation system. FEBS Lett. **579**(10), 2156–2160 (2005)
16. S.H. Hong et al., Cell-free protein synthesis from a release factor 1 deficient *Escherichia coli* activates efficient and multiple site-specific nonstandard amino acid incorporation. Acs Synth. Biol. **3**(6), 398–409 (2014)
17. J.D. Dinman, T. Icho, R.B. Wickner, A-1 ribosomal frameshift in a double-stranded RNA virus of yeast forms a gag-pol fusion protein. Proc. Natl. Acad. Sci. USA **88**(1), 174–178 (1991)
18. T. Hohsaka et al., Incorporation of nonnatural amino acids into proteins by using various four-base codons in an *Escherichia coli* in vitro translation system. Biochemistry **40**(37), 11060–11064 (2001)
19. H. Taira, T. Hohsaka, M. Sisido, In vitro selection of tRNAs for efficient four-base decoding to incorporate non-natural amino acids into proteins in an Escherichia coli cell-free translation system. Nucleic Acids Res. **34**(5), 1653–1662 (2006)
20. T. Hohsaka et al., Efficient incorporation of nonnatural amino acids with large aromatic groups into streptavidin in In Vitro protein synthesizing systems. J. Am. Chem. Soc. **121**(1), 34–40 (1999)

21. T. Hohsaka, M. Fukushima, M. Sisido, Nonnatural mutagenesis in *E. coli* and rabbit reticulocyte lysates by using four-base codons. Nucleic Acids Res. Suppl. (2), 201–202 (2002)
22. R. Krishnakumar, J. Ling, Experimental challenges of sense codon reassignment: an innovative approach to genetic code expansion. FEBS Lett. **588**(3), 383–388 (2014)
23. Y. Shimizu et al., Cell-free translation reconstituted with purified components. Nat. Biotechnol. **19**(8), 751–755 (2001)
24. V.T. Dien et al., Expansion of the genetic code via expansion of the genetic alphabet. Curr. Opin. Chem. Biol. **46**, 196–202 (2018)
25. I. Hirao, M. Kimoto, Unnatural base pair systems toward the expansion of the genetic alphabet in the central dogma. Proc. Jpn. Acad. Ser. B Phys. Biol. Sci. **88**(7), 345–367 (2012)
26. A. Singh-Blom, R.A. Hughes, A.D. Ellington, An amino acid depleted cell-free protein synthesis system for the incorporation of non-canonical amino acid analogs into proteins. J. Biotechnol. **178**, 12–22 (2014)
27. I. Maslennikov, S. Choe, Advances in NMR structures of integral membrane proteins. Curr. Opin. Struct. Biol. **23**(4), 555–562 (2013)
28. M. Neerathilingam et al., Quantitation of protein expression in a cell-free system: efficient detection of yields and 19F NMR to identify folded protein. J. Biomol. NMR **31**(1), 11–19 (2005)
29. D. Kiga et al., An engineered *Escherichia coli* tyrosyl-tRNA synthetase for site-specific incorporation of an unnatural amino acid into proteins in eukaryotic translation and its application in a wheat germ cell-free system. Proc. Natl. Acad. Sci. USA **99**(15), 9715–9720 (2002)
30. K.V. Loscha et al., Multiple-site labeling of proteins with unnatural amino acids. Angew. Chem. Int. Ed. Engl. **51**(9), 2243–2246 (2012)
31. R. Abe et al., Ultra Q-bodies: quench-based antibody probes that utilize dye-dye interactions with enhanced antigen-dependent fluorescence. Sci. Rep. **4**, 4640 (2014)
32. D. Kajihara et al., FRET analysis of protein conformational change through position-specific incorporation of fluorescent amino acids. Nat. Methods **3**(11), 923–929 (2006)
33. H. Nakata, T. Ohtsuki, M. Sisido, A protease inhibitor discovery method using fluorescence correlation spectroscopy with position-specific labeled protein substrates. Anal. Biochem. **390**(2), 121–125 (2009)
34. Y. Ravikumar et al., Unnatural amino acid mutagenesis-based enzyme engineering. Trends Biotechnol. **33**(8), 462–470 (2015)
35. K. Deepankumar et al., Enhancing the biophysical properties of mRFP1 through incorporation of fluoroproline. Biochem. Biophys. Res. Commun. **440**(4), 509–514 (2013)
36. J.R. Kintzing, M.V. Filsinger Interrante, J.R. Cochran, Emerging strategies for developing next-generation protein therapeutics for cancer treatment. Trends Pharmacol. Sci. **37**(12), 993–1008 (2016)
37. Y. Lu et al., Assessing sequence plasticity of a virus-like nanoparticle by evolution toward a versatile scaffold for vaccines and drug delivery. Proc. Natl. Acad. Sci. USA **112**(40), 12360–12365 (2015)
38. G. Yin et al., RF1 attenuation enables efficient non-natural amino acid incorporation for production of homogeneous antibody drug conjugates. Sci. Rep. **7**(1), 3026 (2017)
39. M.H. Rasmussen et al., Pegylated long-acting human growth hormone is well-tolerated in healthy subjects and possesses a potential once-weekly pharmacokinetic and pharmacodynamic treatment profile. J. Clin. Endocrinol. Metab. **95**(7), 3411–3417 (2010)
40. H. Cho et al., Optimized clinical performance of growth hormone with an expanded genetic code. Proc. Natl. Acad. Sci. USA **108**(22), 9060–9065 (2011)
41. J. Mu et al., FGF21 analogs of sustained action enabled by orthogonal biosynthesis demonstrate enhanced antidiabetic pharmacology in rodents. Diabetes **61**(2), 505–512 (2012)

Chapter 4
Cell-Free Biosensing

Biosensing has precise sensitivity and high specificity, and the potential use of biosensor ranges from environmental monitoring, food safety, to disease diagnosis [1]. Biosensing is using the biological sensing element to detect some substances, include inorganic molecules, disease biomarkers, and others. A typical biosensor usually includes biological sensing elements and transducers. The transducer can convert the chemical information generated during the biochemical reaction process into corresponding physical signals, such as optical signals, magnetic signals, and electrical signals. In recent years, with the development of synthetic biology, a large number of cell-based biosensors have been developed, and their constructions become more complicated. Cell-based biosensors have a wide range of detection ability. However, there are still some issues that need to be addressed, like transmembrane transport limitations, the need to maintain cell activity, and a long time to assay [2]. Another problem is that cell-based biosensors make use of living, genetically modified microorganisms (GMOs). The application of these GMO biosensors is often not possible due to biosafety concerns, as it must be prevented that GMOs are released into the environment. To address these limitations of cell-based biosensors, cell-free synthesis system as a platform for biosensing has been developed [3].

Cell-free synthesis systems perform biological transcription and translation activities in vitro. They have a more open environment than cell systems and have no transmembrane transportation limitations. Besides, cell-free systems do not need to maintain cell activity and can operate in a toxic environment that would inhibit or kill cells. More important, cell-free systems do not use GMOs, so they have no biosafety issues and have the potential for practical applications [4]. Researchers have developed cell-free systems to detect some substances, including small molecules, biomarkers, and viruses (Fig. 4.1). In addition to identifying substances, cell-free systems are able to sense some physical signals, like the light.

© The Author(s), under exclusive license to Springer Nature Singapore Pte Ltd. 2020   21
Y. Lu, *Cell-Free Synthetic Biology*, SpringerBriefs in Applied Sciences
and Technology, https://doi.org/10.1007/978-981-13-1171-0_4

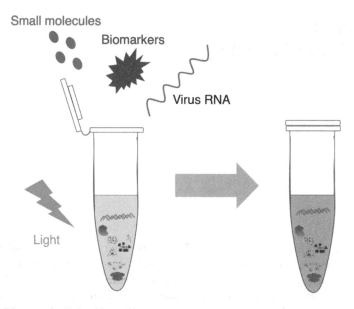

Fig. 4.1 Diagram of cell-free biosensing

4.1 Biomarker Detection

Biomarkers are generally related to cell growth and proliferation, and therefore the state of the human body can be known by measuring them. Detecting a disease-specific biomarker may help in the identification, early diagnosis and prevention of disease, and on-time monitoring during treatment. Acyl-homoserine lactone (AHL) was a type of quorum sensing signal molecule that regulates the expression of many physiological properties. N-butyryl-homoserine lactone (C4-HSL) and N-3-oxo-dodecanoyl-homoserine lactone (3OC12-HSL), produced by *Pseudomonas aeruginos*, were present in the sputum, urine, and blood of many patients with cystic fibrosis [5], which show clinical relevance with links to infection state [6]. Therefore, it is important to establish an effective approach to detecting AHLs.

The cell-free biosystem has been successfully used for biosensing QS molecules. Wen et al. detected QS molecule 3OC12-HSL at nanomolar levels by cell-free biosystem [7]. They constructed a biosensor that responded to 3OC12-HSL and tested the specificity of the sensor in a cell-free system. The response of different cell extracts was compared. Cell-free biosensing system with *E. coli* Rosetta as cell extract has better performance. 3OC12-HSL from the sputum of patients with CF was extracted and tested in the cell-free system, and the results were compared with LC-MS measurements.

The cell-free system generally uses well-characterized transcription factor ligands to detect biomarkers. However, many biomarkers do not have corresponding transcription factor ligand, which significantly limits the detection of biomarkers.

Voyvodic et al. proposed a strategy, adding transformed protein to the cell-free system that converts a biomarker without ligand protein into another biomarker, which has ligand protein [8]. They added HipO and CocE enzymes to the cell-free system to convert hippuric acid and cocaine to benzoic acid. It allows the cell-free system for detecting hippuric acid and cocaine and expands the cell-free detection range. This may be another way to use cell-free systems to identify biomarkers.

4.2 Virus Detection

The virus is composed of a nucleic acid molecule (DNA or RNA) and protein, which is extremely harmful to human. The emerging outbreak of Ebola and Zika viruses has threatened the health of people around the world, which resulted in the World Health Organization (WHO) to call for the rapid development of virus diagnostics [9]. To reduce the cost and time of virus detection, synthetic biologists have developed two biotechnologies. The first technology is programmable RNA sensors called toehold switches. Any RNA sequence can be detected by rational design. The second technology is called paper-based cell-free system, which allows cell-free synthesis systems to be lyophilized on paper for biosensing outside the laboratory [10]. They combined these two technologies to create a low-cost, fast platform for virus detection.

The workflow for detecting Zika virus by using sequence-specific method can be divided into three parts [11]. First, design primers of isothermal RNA amplification and toehold switch RNA sensors based on the information from online databases. Then, the synthetic primers are used to construct the toehold switch RNA sensors and verify that it is capable of detecting the corresponding RNA sequence. Finally, the validated sensor and cell-free system are embedded in the paper for freeze-drying. When a diagnostic test is required, the extracted RNA is subjected to isothermal amplification, and the amplified RNA solution is dropped onto the test strip. If the test strip changes from yellow to purple, the sample contains the responding viral RNA. By this detection method, the Zika virus can be detected without being interfered by other viruses closely related to the Zika virus. It is possible to distinguish the Zika virus belong to different regions by introducing CRISPR/Cas9-based module in cell-free systems.

Cell-free systems are easy to perform molecular design due to the open environment, and the cell-free system can be frozen on paper, extending its stability and increasing portability. The use of cell-free systems for viral diagnosis is an essential direction for the future development of cell-free biosensing.

4.3 Light Detection

It is difficult to program a spatiotemporal control of gene expression by a more straightforward way in the cell-free system [12]. This required us to understand the structural changes of enzymes and temporal variations of gene expression in genetic networks and metabolic pathways outside a living cell. Regardless, an ideal programmable dynamic control tool is critical to achieving this challenge. As an ideal control switch, light has the advantages of fast response time, good space-time conversion and non-toxic and harmless. Some researchers have developed photo-sensitive proteins to control gene expression in *E. coli* [13]. These provide a good basis for cell-free light-controlled gene expression. Jayaraman et al. achieved a controlled expression of a red fluorescent protein in a cell-free system by purifying the EL222 photosensitizing protein and adding it to a commercial cell-free kit [14]. The difference in expression between light and dark is 10 times. Next, they construct a plasmid including EL222 protein and a fluorescent protein, through computer simulation optimization. The differences in expression between light and dark finally reached 5 times.

Most light-control genetic systems require two proteins, one photosensitive protein, and the other corresponding regulatory protein. And a light-controlled gene expression system involves the synergy of two proteins. It remains challenging to realize the interaction of multiple proteins in a cell-free system. Besides, a variety of different color of light for controlling the gene expression has been achieved in the cell system. Therefore, there would be still a lot of work to be explored in cell-free light-control gene expression.

References

1. A.P. Turner, Biosensors: sense and sensibility. Chem. Soc. Rev. **42**(8), 3184–3196 (2013)
2. Q. Liu et al., Cell-based biosensors and their application in biomedicine. Chem. Rev. **114**(12), 6423–6461 (2014)
3. M.T. Smith et al., The emerging age of cell-free synthetic biology. FEBS Lett. **588**(17), 2755–2761 (2014)
4. T. Pellinen, T. Huovinen, M. Karp, A cell-free biosensor for the detection of transcriptional inducers using firefly luciferase as a reporter. Anal. Biochem. **330**(1), 52–57 (2004)
5. C.E. Chambers et al., Identification of N-acylhomoserine lactones in mucopurulent respiratory secretions from cystic fibrosis patients. FEMS Microbiol. Lett. **244**(2), 297–304 (2005)
6. H.L. Barr et al., Diagnostic and prognostic significance of systemic alkyl quinolones for *P. aeruginosa* in cystic fibrosis: a longitudinal study; response to comments. J. Cyst. Fibros. **16**(6), e21 (2017)
7. K.Y. Wen et al., A cell-free biosensor for detecting quorum sensing molecules in *P. aeruginosa*-infected respiratory samples. ACS Synth. Biol. **6**(12), 2293–2301 (2017)
8. P.L. Voyvodic et al., *Plug-and-Play Metabolic Transducers Expand the Chemical Detection Space of Cell-Free Biosensors* (2018)
9. D.W. Smith, J. Mackenzie, Zika virus and Guillain-Barre syndrome: another viral cause to add to the list. Lancet **387**(10027), 1486–1488 (2016)
10. K. Pardee et al., Paper-based synthetic gene networks. Cell **159**(4), 940–954 (2014)

11. K. Pardee et al., Rapid, low-cost detection of Zika virus using programmable biomolecular components. Cell **165**(5), 1255–1266 (2016)
12. H. Niederholtmeyer, V. Stepanova, S.J. Maerkl, Implementation of cell-free biological networks at steady state. Proc. Natl. Acad. Sci. USA **110**(40), 15985–15990 (2013)
13. J.J. Tabor, A. Levskaya, C.A. Voigt, Multichromatic control of gene expression in *Escherichia coli*. J. Mol. Biol. **405**(2), 315–324 (2011)
14. P. Jayaraman et al., Cell-free optogenetic gene expression system. ACS Synth. Biol. **7**(4), 986–994 (2018)

Chapter 5
Artificial Life

5.1 Artificial Cell Construction

Cells are considered to be "life blocks" as the basic structure and functional unit of living things [1]. Research on cell biology has not been limited to its structural function. Many new fields, such as drug delivery, biosensors, bioremediation, drug preparation, and the origin of life, require broad studies on cells. However, with the development of cell biology, some shortcomings such as complexity and vulnerability of cells affect people's exploration of new research fields. To solve these problems, artificial cells have been constructed to simulate biological cells. Artificial cells are controllable and more robust than natural cells [2].

At present, the design and construction of artificial cells are being carried out in two ways: the reduction of the genome in the bacteria from top to bottom and the "bottom-up" synthesis and integration of DNA, RNA, protein, and membrane in vitro (Fig. 5.1) [3]. The top-down approach is to sequence the genome, identify non-essential genes, remove the non-essential genes from the genome by oligonu-cleotide synthesis, and finally transplant the genome into the appropriate recipient cells to synthesize artificial cells [4]. The "bottom-up" approach involves synthesis and self-replication by combining the necessary biomacromolecules, genes, and small molecule substrates. This method removes genes of unknown function and makes it easier to manipulate and adjust the systems flexibly [5].

Cell-free biosystems have been well developed for the construction of artificial cells in a bottom-up way. There are three indispensable parts for constructing artificial cells from the bottom up, namely cell compartment (cell membrane), synthesis device (energy processing and regeneration), and information molecules (RNA or DNA) [6]. The construction of a stable membrane is critical to the proper functioning of artificial cells. As a protective shell, the membrane provides a restrictive boundary for artificial cells, so that it is not interfered with by the extracellular environment during biosynthesis. Information molecules define the nature and function of cells. Transcription and translation are the key processes or basis for living metabolism.

© The Author(s), under exclusive license to Springer Nature Singapore Pte Ltd. 2020
Y. Lu, *Cell-Free Synthetic Biology*, SpringerBriefs in Applied Sciences
and Technology, https://doi.org/10.1007/978-981-13-1171-0_5

Fig. 5.1 Approaches for the design and construction of artificial cells

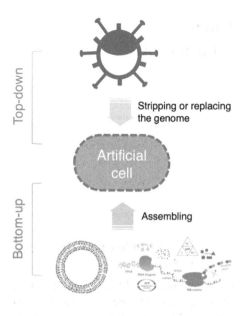

Modern cell-free biosystems have been developed as a research tool to perform transcription and translation.

The cell-free system based artificial cells has been built for studying biological functions, revealing the origins of life, or healthcare applications. In 2011, Mei Li and David C. Green created a primitive bio-inorganic cell model. This model is structurally composed of a porous inorganic membrane [7]. Also, they found that when the cell-free system is encapsulated in the colloidosome interior, it expresses substantially the same green fluorescent protein as the solution system. They reported considerable enhancements in the specific activity of enzymes when entrapped within the nanoparticle-stabilized water droplets. The results suggested that artificial cells had considerable potential in synthetic biology and bionanotechnology. Two vesicles were constructed as substrates for artificial cells. These vesicles can be used to separate different biological processes [8]. Compartment vesicles were generated using the phase transfer of water-in-oil droplets, which contained expression plasmids and cell-free biosynthesis systems. Vesicle artificial cells allow spatial separation of complex biological processes. This discovery can be widely used in microreactors and medical environments. Ho et al. described a method of constructing artificial cells [9]. This method can simulate more biological cell behavior. The microfluidic double emulsion system is used to encapsulate the mammalian cell-free expression system, which can express membrane proteins into bilayer or soluble proteins in vesicles. Krinsky et al. constructed artificial lipid vesicles [10]. The vesicles can be used to synthesize anticancer proteins by encapsulating the cell-free system. Synthetic cells are designed to be independent systems that harvest nutrients from their biological microenvironment to trigger protein synthesis. It may be a new platform for on-demand synthesis of therapeutic proteins for treating diseases.

5.2 Bacteriophage Synthesis

Phage is a virus that invades bacteria. It is also a genetic material that endows the biological characteristics of host bacteria. Phage must be parasitic in living bacteria with strict host specificity. Specificity depends on the molecular structure and complementarity of the receptors on the surface of organs and receptors adsorbed by phages. Phage is the most common and widely distributed group of viruses. In 1915, the bacteriophage was discovered and used as an antimicrobial agent [11]. As phage has been found to be unstable, controversy has arisen about the effectiveness and repeatability of the use of phages in the treatment of bacterial infections [12, 13]. Due to the significant diversity of bacteriophages in structure, life cycle and genome organization, obtaining regulatory approval for the use of such cocktails may be challenging [14]. Phage can lead to rapid and large-scale bacterial lysis. This is likely to induce adverse immune responses in human hosts [15]. But as bacteria evolve, they can avoid infection by phages.

In order to construct functional phage particles in vitro, researchers need to reintroduce phage DNA into host bacteria by transformation. However, this process requires high transformation efficiency, which is also a key bottleneck of modern bacteriophage genetic engineering.

Cell-free synthetic biology offers a potential solution to these challenges. Cell-free biosystems have been successfully used to replicate, synthesize, and assemble the phage T7, ΦX174, and MS2 [16, 17] (Fig. 5.2). After several hours of incubation, more than one billion infectious bacteriophages are produced per milliliter of reaction. Viral DNA of genome size can be expressed in test tubes, and the whole information processing chain, including replication of DNA instructions, is redefined. This provides new possibilities for programming and studying complex biochemical systems in vitro.

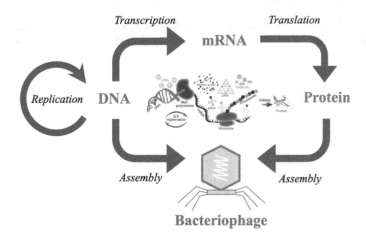

Fig. 5.2 Cell-free biosynthesis of bacteriophage

References

1. B.R. Masters, *History of the Electron Microscope in Cell Biology* (2009)
2. A. Pohorille, D. Deamer, Artificial cells: prospects for biotechnology. Trends Biotechnol. **20**(3), 123–128 (2002)
3. P.L. Luisi, P. Walde, T. Oberholzer, Lipid vesicles as possible intermediates in the origin of life. Curr. Opin. Colloid Interface Sci. **4**(1), 33–39 (1999)
4. D.G. Gibson, J.C. Venter, Creation of a bacterial cell controlled by a chemically synthesized genome. Science **329**(5987), 52 (2010)
5. E.M. Evans, Cognitive and contextual factors in the emergence of diverse belief systems: creation versus evolution. Cogn. Psychol. **42**(3), 217–266 (2001)
6. V. Noireaux, Y.T. Maeda, A. Libchaber, Development of an artificial cell, from self-organization to computation and self-reproduction. Proc. Natl. Acad. Sci. USA **108**(9), 3473–3480 (2011)
7. M. Li et al., In vitro gene expression and enzyme catalysis in bio-inorganic protocells. Chem. Sci. **2**(9), 1739–1745 (2011)
8. Y. Elani, R.V. Law, O. Ces, Protein synthesis in artificial cells: using compartmentalisation for spatial organisation in vesicle bioreactors. Phys. Chem. Chem. Phys. **17**(24), 15534–15537 (2015)
9. K.K. Ho, V.L. Murray, A.P. Liu, Engineering artificial cells by combining HeLa-based cell-free expression and ultrathin double emulsion template. Methods Cell Biol. **128**, 303–318 (2015)
10. N. Krinsky et al., Synthetic cells synthesize therapeutic proteins inside tumors. Adv. Healthc. Mater. **7**(9), e1701163 (2018)
11. F.W. Twort, An investigation on the nature of ultra-microscopic viruses. The Lancet **186**(4814), 1241–1243 (1915)
12. P. Caumette, J.C. Bertrand, P. Normand, *Some Historical Elements of Microbial Ecology* (Springer Netherlands, 2015), pp. 9–24
13. W. Xavier, D.R. Sophie, S.M. Opal, A historical overview of bacteriophage therapy as an alternative to antibiotics for the treatment of bacterial pathogens. Virulence **5**(1), 226–235 (2014)
14. C. Loccarrillo, S.T. Abedon, Pros and cons of phage therapy. Bacteriophage **1**(2), 111–114 (2011)
15. A. Górski et al., Phage as a modulator of immune responses: practical implications for phage therapy. Adv. Virus Res. **83**, 41–71 (2012)
16. J. Garamella et al., The all *E. coli* TX-TL toolbox 2.0: a platform for cell-free synthetic biology. Acs Synth. Biol. **5**(4), 344–355 (2016)
17. J. Shin, P. Jardine, V. Noireaux, Genome Replication, Synthesis, and Assembly of the Bacteriophage T7 in a Single Cell-Free Reaction. Acs Synth. Biol. **1**(9), 408–413 (2012)

Chapter 6
Other Emerging Development Trends

6.1 Post-translational Modification

Many proteins require post-translational modification (PTM) to maintain their biological activity. PTMs can significantly affect overall protein characteristics, including stability and solubility. In addition to changing the physical, chemical and structural characteristics of amino acid sequences, PTM is also a major determinant of successful protein synthesis. It should be noted that many eukaryotic proteins require multiple PTMs to achieve natural and bioactive conformation [1]. Although cell-free biosystems have been developed as a robust protein engineering and synthesis platform, one of its disadvantages is that PTM is not as good as cellular systems. To address these challenges, some cell-free approaches have been developed.

Glycosylation is one of the major PTMs. N-glycosylation is necessary for proper folding of proteins, as more than half of eukaryotic proteins are predicted to be glycoproteins [2]. To perform adequate glycosylation, mammalian cell-based cell-free biosystems have been developed to use the native glycosylation pathways. Yadavalli et al. successfully expressed the glycosylated malaria proteins in one step or two steps for vaccine applications by HeLa cell-based cell-free expression system [3]. The expressed malaria proteins in the cell-free reaction system could be readily purified by the affinity purification approach and used for sequencing, immunoassay, and immunization. Then prokaryotic cell-based cell-free systems are also developed for the production of homogeneous glycoproteins. Jaroentomeechai et al. proposed a novel glycoprotein synthesis technique using a cell-free system combined with asparagine-linked glycosylation [4]. The method utilizes a glyco-optimized *E. coli* strain to obtain a cell extract with glycosylation components, including lipid-linked oligosaccharides and oligosaccharyltransferases. The efficient synthesis of glycoproteins with site-specific glycosylation not only help better understand the PTM process but also make on-demand biopharmaceuticals.

Protein phosphorylation is a process in which phosphate groups reversibly attach to proteins. Protein phosphorylation is a crucial PTM in nature and shows an essential mechanism to identify protein functions and molecular diversity [5]. Among them,

© The Author(s), under exclusive license to Springer Nature Singapore Pte Ltd. 2020
Y. Lu, *Cell-Free Synthetic Biology*, SpringerBriefs in Applied Sciences
and Technology, https://doi.org/10.1007/978-981-13-1171-0_6

serine phosphorylation is a relatively abundant PTM, which can affect many key cell processes, including cell metabolism and signal transduction. An effective method of phosphorylation is the site-specific binding of phosphoserine to proteins, whose studies can help answer many basic biological questions [6]. Nemoto et al. used a wheat germ cell-free biosynthesis system and a luminescence system to analyze the threonine/serine autophosphorylation activity of protein kinases from the plant in a high-throughput format. This study can help find many unknown protein kinases involved in phosphor-signaling pathways and understand their function for practical applications [7]. Oza et al. developed a cell-free protein synthesis system based on the chassis *E. coli* strain, which possessed site-specific phosphoserine binding activity and was used for the synthesis of human MEK1 kinase. The results demonstrated that MEK1 was expressed with phosphorylation, and the monophosphorylation event was sufficient for the activation of MEK1 [8].

6.2   Biomaterials

Nowadays, only a small amount and categories of biomaterials are made by microorganism like bacteria, and most of them are protein-based. Cell-free protein synthesis system now can be used as an even novel system to produce biomaterials. The cell-free system has a much higher tolerance of substances which used to be toxic to cells. And also, in the cell-free system, there is no space limit for biomaterials synthesis and assembly. Cell-free biosystems have been well used for the synthesis and assembly of virus-like particles (VLP) [9, 10]. VLPs can be used as drug delivery vesicle and building blocks for biomaterials. Compared with cellular systems, taking advantage of the cell-free system to produce VLPs is simpler and highly controllable, with the VLPs assembly conditions being manipulated in ways which are difficult to reach in cell systems. Cell-free methods can be used as a convenient tool to design self-assembled proteins [11]. The projection of the assembled micro-scale array is consistent with the design model and shows the symmetry of the target layer group. The programmable microscale two-dimensional protein array synthesized in the cell-free system could provide new methods for structural biology studies and nanoengineering. It has always been a considerable obstacle to produce large size proteins in vivo. Some will be misfolded and become inclusion bodies, having low solubility in the solution. Researchers have proved that cell-free systems can be widely used in production, solving many problems protein production in cell culture has faced. By manipulating the cell-free reaction conditions, optimization of yield, solubility or even stability is not impossible anymore. Because of this, cell-free biosystems is a powerful and attractive tool for designing and producing novel biomaterials.

6.3 Metabolic Engineering

The use of metabolic engineering to produce small molecule chemicals through biosynthesis of microbial metabolism has become increasingly popular. For example, biosynthesis can be achieved by introducing a metabolic pathway into a model microorganism to produce 1,3-propanediol, farnesene, artemisinin, etc. However, the use of microbial fermentation in vivo for design-build-test (DBT) research cycles is often time-consuming [12]. How to speed up the DBT cycle is an essential issue in the field of metabolic engineering. The new cell-free system is used for prototyping rapidly in vitro metabolic pathways. In addition, due to its openness, the cell-free reaction system has the following advantages [12–14]: it is easier to control the reaction system precisely; the control of cell extracts enables the associated metabolic processes to be activated in vitro; it is more capable of concentrating all components for synthesis on specific proteins or small molecule products. For example, using the advantages of cell-free synthesis, the method was applied to the synthesis of n-butanol. It suggests that the *E. coli*-based cell-free system can complete 17-step n-butanol metabolic pathways in vitro with high metabolic activity. The open cell-free system directly regulates the physicochemical conditions of the reaction, which helps the in-depth study of the active site of the enzyme, discovery of new enzymes, and the prototyping of new metabolic pathways. Compared with the fermentation process of the microorganisms, the cell-free approach presents a faster iteration that shortens the research cycle [12].

6.4 Genomics Analysis

Cell-free biosystem is also widely used in genomics analysis because of its high-throughput, rapid response and in vitro synthesis. Compared with the time-consuming culture expression and complex metabolic network of the cell system, cell-free biosystem provides more convenient performance evaluation of gene expression regulatory elements (promoter, transcription factor, etc.), and the open cell-free system can also be controlled by variable parameters to explore resource constraints and competitive relationships to deepen understanding of cellular transcription and translation processes. The cell-free biosystem mimics the intracellular characteristics of the genome, and the in vitro genomic transcription-translation system developed by using different model microorganisms and genomes demonstrates the de novo synthesis of RNA and proteins, whose activity is regulated by the genomic structure and gene location in the genome [15]. By using cell-free biosystem, rapid acquirement and model-based analysis could be got from nonmodel bacteria *Bacillus megaterium*. Cell-free biosystems were used to measure and model uncharacterized endogenous constitutive and inducible promoters previously, ribosome binding site variants, and previously unknown transcription factor binding affinity [16].

6.5   Genome Editing

As a disruptive technology in genetic engineering, CRISPR technology has proven to be widely used genome editing methods for versatile biological applications. However, the implementation of CRISPR-Cas systems has been surpassed by the continued discovery of new anti-CRISPR proteins and Cas nucleases. Furthermore, the current process of characterizing the basic properties and functions of these proteins is lengthy and cumbersome, thus limiting their development. Meanwhile, performance is reduced when testing large amounts of protein or gRNA simultaneously. However, cell-free biosystem provides an excellent solution to this problem and is a fast and scalable method of characterization that greatly facilitates the characterization and application of CRISPR technology in its many forms. Using the *E. coli* cell-free biosystem to express the active CRISPR mechanism in vitro [17], measuring the gene suppression dynamics and the DNA cleavage of single-cut and multi-effect CRISPR nucleases, and determining the specificity of different anti-CRISPR proteins to rapid development of scalable screening for protospacer adjacent motifs has been successfully applied to five non-specific Cpf1 nucleases. Cell-free biosynthesis platform can be a powerful platform to discover, screen, and design novel genome editing tools.

6.6   Microfluidics

Combining cell-free systems with microfluidic control technology is also a hot topic of research, which can blend the advantages of both and complement each other. Although cell-free biosynthesis system has the benefits of openness and easy regulation, the short-lived and high reagent cost still limit its further development, and the new automatic microfluidics technology can achieve continuous, controllable and resource saving process. For example, the microfluidic platform addresses individual biochemical analysis requirements of the transcription/translation steps of cell-free systems in different compartments and combines different reaction steps by quasi-continuous transfer of gene templates. The immobilization of specific RNA templates demonstrates transcriptional compatibility and reusability of immobilized particles, making automated modular cell-free bioreactors possible [18].

Furthermore, the microfluidic platform can further optimize and supplement the cell-free biosystem, such as enhancing the visualization of the cell-free reaction process, allowing us to understand the molecular assembly process better, and can be more detailed for specific point-of-care, which can help make cell-free technology more powerful. In situ cell-free protein synthesis, assembly and TEM imaging on a microfluidic platform represented by biochips, resulting in a protein space pattern assembled on the surface, allowing for more intuitive exploration of the relevant reaction mechanisms, seeing many exciting phenomena [19]. The optimization of the point-of-care can promote the synthesis of single-dose therapeutic protein, and the parallel material exchange of "reactor" and "feeder" can be realized through the design of the serpentine channel bioreactor combined with the nano-membrane materials, thus extending the reaction time and significantly increasing the protein yield [20].

References

1. A.A. Tokmakov et al., Multiple post-translational modifications affect heterologous protein synthesis. J. Biol. Chem. **287**(32), 27106–27116 (2012)
2. A. Helenius, M. Aebi, Roles of N-linked glycans in the endoplasmic reticulum. Annu. Rev. Biochem. **73**, 1019–1049 (2004)
3. R. Yadavalli, T. Sam-Yellowe, HeLa based cell free expression systems for expression of plasmodium rhoptry proteins. J. Vis. Exp. **100**, e52772 (2015)
4. T. Jaroentomeechai et al., Single-pot glycoprotein biosynthesis using a cell-free transcription-translation system enriched with glycosylation machinery. Nat. Commun. **9**(1), 2686 (2018)
5. C.S. Tan et al., Comparative analysis reveals conserved protein phosphorylation networks implicated in multiple diseases. Sci. Signal **2**(81), 39 (2009)
6. S. Lee et al., A facile strategy for selective incorporation of phosphoserine into histones. Angew. Chem. Int. Ed. Engl. **52**(22), 5771–5775 (2013)
7. K. Nemoto et al., Autophosphorylation profiling of Arabidopsis protein kinases using the cell-free system. Phytochemistry **72**(10), 1136–1144 (2011)
8. J.P. Oza et al., Robust production of recombinant phosphoproteins using cell-free protein synthesis. Nat. Commun. **6** (2015)
9. Y. Lu et al., Assessing sequence plasticity of a virus-like nanoparticle by evolution toward a versatile scaffold for vaccines and drug delivery. Proc. Natl. Acad. Sci. USA **112**(40), 12360–12365 (2015)
10. Y. Lu et al., *Escherichia coli*-based cell free production of flagellin and ordered flagellin display on virus-like particles. Biotechnol. Bioeng. **110**(8), 2073–2085 (2013)
11. S. Gonen et al., Design of ordered two-dimensional arrays mediated by noncovalent protein-protein interfaces. Science **348**(6241), 1365–1368 (2015)
12. A.S. Karim, M.C. Jewett, A cell-free framework for rapid biosynthetic pathway prototyping and enzyme discovery. Metab. Eng. **36**, 116–126 (2016)
13. A.S. Karim, M.C. Jewett, Cell-free synthetic biology for pathway prototyping. Methods Enzymol. **608**, 31–57 (2018)
14. J.R. Swartz, Expanding biological applications using cell-free metabolic engineering: an overview. Metab. Eng. (2018)
15. K. Fujiwara et al., In vitro transcription-translation using bacterial genome as a template to reconstitute intracellular profile. Nucleic Acids Res. **45**(19), 11449–11458 (2017)
16. S.J. Moore et al., Rapid acquisition and model-based analysis of cell-free transcription-translation reactions from non-model bacteria. Proc. Natl. Acad. Sci. USA **115**(19), E4340–E4349 (2018)
17. R. Marshall et al., Rapid and scalable characterization of CRISPR technologies using an *E. coli* cell-free transcription-translation system. Mol. Cell **69**(1), 146–157 e3 (2018)
18. V. Georgi et al., On-chip automation of cell-free protein synthesis: new opportunities due to a novel reaction mode. Lab Chip **16**(2), 269–281 (2016)
19. Y. Heyman et al., Cell-free protein synthesis and assembly on a biochip. Nat. Nanotechnol. **7**(6), 374–378 (2012)
20. A.C. Timm et al., Toward microfluidic reactors for cell-free protein synthesis at the point-of-care. Small **12**(6), 810–817 (2016)

Chapter 7
Conclusions

To address challenges cell-based synthetic biology faces, CFSB has been fast developed as a robust enabling technology to engineer biological parts and networks for versatile applications, such as biopharmaceuticals, medical diagnostics, and others. Without using living cells, the biological transcription, translation, metabolism, and network can be engineered with unprecedented freedom in defined or undefined cell-free biosystems. Cell-free biosystems now can act as powerful testing platforms for better understanding biology and speeding up the design-build-test research cycle, and also serve as flexible biomanufacturing platforms for the synthesis of natural and unnatural proteins on demand. Because of the open nature, cell-free biosystems can work seamlessly together with other fast-growing advanced technologies, such as materials science. In the future development, more defined or undefined cell-free biosystems based on different kinds of prokaryotic or eukaryotic cells need to be constructed to fulfill different needs of biopharmaceutical protein synthesis. Furthermore, the organization of the cell-free reaction process needs to consider the spatial effect for better protein synthesis. CFSB will be popular mainly for fast prototyping or testing, by integrating with other next-generation biotechnologies. To realize the full potential of CFSB, the primary focus of CFSB studies will be doing what cell cannot or is hard to do.

© The Author(s), under exclusive license to Springer Nature Singapore Pte Ltd. 2020   37
Y. Lu, *Cell-Free Synthetic Biology*, SpringerBriefs in Applied Sciences
and Technology, https://doi.org/10.1007/978-981-13-1171-0_7

9789811311703